YOUR KNOWLEDGE HAS VALUE

- We will publish your bachelor's and master's thesis, essays and papers

- Your own eBook and book - sold worldwide in all relevant shops

- Earn money with each sale

Upload your text at www.GRIN.com
and publish for free

Bibliographic information published by the German National Library:

The German National Library lists this publication in the National Bibliography; detailed bibliographic data are available on the Internet at http://dnb.dnb.de .

This book is copyright material and must not be copied, reproduced, transferred, distributed, leased, licensed or publicly performed or used in any way except as specifically permitted in writing by the publishers, as allowed under the terms and conditions under which it was purchased or as strictly permitted by applicable copyright law. Any unauthorized distribution or use of this text may be a direct infringement of the author s and publisher s rights and those responsible may be liable in law accordingly.

Imprint:

Copyright © 2018 GRIN Verlag
Print and binding: Books on Demand GmbH, Norderstedt Germany
ISBN: 9783668855953

This book at GRIN:

https://www.grin.com/document/453733

Deapon Biswas

The "Biswas Series". An Arithmetic Series

GRIN Verlag

GRIN - Your knowledge has value

Since its foundation in 1998, GRIN has specialized in publishing academic texts by students, college teachers and other academics as e-book and printed book. The website www.grin.com is an ideal platform for presenting term papers, final papers, scientific essays, dissertations and specialist books.

Visit us on the internet:

http://www.grin.com/

http://www.facebook.com/grincom

http://www.twitter.com/grin_com

The Biswas Series

Deapon Biswas
Transport Officer, Private Concern, Chittagong, Bangladesh.

Abstract

We are familiar with series in high school algebra. Here I introduce a more extensive series with manner due to arithmetic series. The series proposed here have two constant parameters, the number of terms N and its step M, where M ≤ N.

Key Words

Biswas series, B first combination rule, B second combination rule, B third combination rule and B fourth combination rule.

Article Outline

1. Introduction
2. Biswas series
3. B first combination rule
4. B second combination rule
5. B third combination rule
6. B fourth combination rule
7. Conclusions

1. Introduction

It is a finite series of two constant parameters. In this paper I develop 9 theorems to cover this paper and 4 combination rules established as theorems.

2. Biswas series

We have already met the usual series
$$2 + 2 + 2 + 2 + 2 \qquad (1)$$
$$1 + 3 + 5 + 7 + 9 \qquad (2)$$

But it is sometimes requirable more extensive form of these series. We notified that the series (1) is specified with one constant term viz. 2. We may say the series is of step 1 with 5 terms. The series (2) is specified with two constant terms viz. 1 and 2. We may say the series is of step 2 with 5 terms. The steps are 1 and 2. Now taking $1 = V_1$ and $2 = V_2$ we get (2) in the form

$$V_1 + (V_1 + V_2) + (V_1 + 2V_2) + (V_1 + 3V_2) + (V_1 + 4V_2) \qquad (3)$$

Now what happens when a series of step 3 of 5 terms? Because of the foregoing manner we get

$$V_1 + (V_1 + V_2) + (V_1 + 2V_2 + V_3) + (V_1 + 3V_2 + 3V_3) + (V_1 + 4V_2 + 6V_3) \qquad (4)$$

For more convenient, if we interested in finding the mechanical structure of (4) we get the fifth term as

$$(V_1 + 4V_2 + 6V_3) = V_1 \left\{ C\binom{5}{1} - C\binom{4}{1} \right\} + V_2 \left\{ C\binom{5}{2} - C\binom{4}{2} \right\}$$
$$+ V_3 \left\{ C\binom{5}{3} - C\binom{4}{3} \right\}$$

Continuing this process and using induction we can write the Nth term of M step series.

$$= V_1 \left\{ C\binom{N}{1} - C\binom{N-1}{1} \right\} + V_2 \left\{ C\binom{N}{2} - C\binom{N-1}{2} \right\}$$
$$+ V_3 \left\{ C\binom{N}{3} - C\binom{N-1}{3} \right\} + \ldots + V_M \left\{ C\binom{N}{M} - C\binom{N-1}{M} \right\} ; M \leq N$$
$$= \sum_{j=1}^{M} \left[V_j \left\{ C\binom{N}{j} - C\binom{N-1}{j} \right\} \right]$$
$$= \sum_{j=1}^{M} V_j C\binom{N-1}{j-1} \quad \text{[Using Pascal's rule]} \qquad (5)$$

Now we are ready to develop a general form of M step series with N terms as

Theorem 1: Let $V_1, V_2, V_3, \ldots, V_M$ are M steps and V_j, $j = 1, 2, 3, \ldots, M$ are real numbers, then the Biswas series of M steps with N terms denoted by B_M^N; $M \leq N$ is given by

$$B_M^N = V_1 + \sum_{j=1}^{2} V_j C\binom{2-1}{j-1} + \sum_{j=1}^{3} V_j C\binom{3-1}{j-1} + \cdots$$
$$+ \sum_{j=1}^{M} V_j C\binom{M-1}{j-1} + \cdots + \sum_{j=1}^{M} V_j C\binom{N-1}{j-1}; M \leq N \quad \text{---(6)}$$

If you should intuitive to your mind carefully you must lead to state the following statements.

(i) The Biswas series of M step with N terms where $M \leq N$ is always exist. If $M > N$ the series does not exist.

(ii) The Mth step first added in M^{th} term.

(iii) The Nth term is expressed as in terms of M steps viz. $V_1, V_2, V_3, \ldots, V_M$ with M definite coefficients i.e., the Nth terms of a B series B_M^N is written as

$$V_1 C\binom{N-1}{0} + V_2 C\binom{N-1}{1} + V_3 C\binom{N-1}{2} + \cdots + V_M C\binom{N-1}{M-1}$$

Example 1: Let $V_1 = 2$, $V_2 = \frac{1}{3}$ and $V_3 = \frac{3}{4}$ then

(i) find B_3^5

(ii) find 10th term

(iii) show that 3rd step first added in 3rd term.

Solution:

(i) We get

$$B_3^5 = 2 + \sum_{j=1}^{2} V_j C\binom{2-1}{j-1} + \sum_{j=1}^{3} V_j C\binom{3-1}{j-1} + \sum_{j=1}^{3} V_j C\binom{4-1}{j-1}$$
$$+ \sum_{j=1}^{3} V_j C\binom{5-1}{j-1}$$
$$= 2 + \left\{V_1 C\binom{1}{0} + V_2 C\binom{1}{1}\right\} + \left\{V_1 C\binom{2}{0} + V_2 C\binom{2}{1} + V_3 C\binom{2}{2}\right\}$$
$$+ \left\{V_1 C\binom{3}{0} + V_2 C\binom{3}{1} + V_3 C\binom{3}{2}\right\} + \left\{V_1 C\binom{4}{0} + V_2 C\binom{4}{1} + V_3 C\binom{4}{2}\right\}$$
$$= 2 + \frac{7}{3} + \frac{41}{12} + \frac{21}{4} + \frac{47}{6}$$

(ii) We get the 10th terms as

$$\sum_{j=1}^{3} V_j C\binom{9}{j-1} = V_1 C\binom{9}{0} + V_2 C\binom{9}{1} + V_3 C\binom{9}{2}$$
$$= 2 \times 1 + \frac{1}{3} \times 9 + \frac{3}{4} \times 36 = 2 + 3 + 9 = 14.$$

(iii) We found in (i) the 3rd step first added in 3rd term.

3. B first combination rule

Theorem 2: The number of combinations of N different things taken j at a time can be expressed as

$$C\binom{N}{j} = \sum_{i=j}^{N} C\binom{i-1}{j-1} \; ; \; j \leq N \qquad (7)$$

Proof: We get from special combination series

$$C\binom{M}{V} = \sum_{i=V-1}^{M-1} C\binom{i}{V-1}$$

Now taking $M = N$, $V = j$, we get

$$C\binom{N}{j} = \sum_{i=j-1}^{N-1} C\binom{i}{j-1} = \sum_{i=j}^{N} C\binom{i-1}{j-1}$$

Example 2: Expand (i) $C\binom{5}{4}$ and (ii) $C\binom{7}{3}$

Solution: (i) $C\binom{5}{4} = \sum_{i=4}^{5} C\binom{i-1}{j-1} = C\binom{3}{3} + C\binom{4}{3} = 1 + 4 = 5 = C\binom{5}{4}$.

Thus, L. H. S. = R. H. S.

(ii) $C\binom{7}{3} = \sum_{i=3}^{7} C\binom{i-1}{j-1} = C\binom{2}{2} + C\binom{3}{2} + C\binom{4}{2} + C\binom{5}{2} + C\binom{6}{2}$

$= 1 + 3 + 6 + 10 + 15 = 35 = C\binom{7}{3}$

Thus, L. H. S. = R. H. S.

4. B second combination rule

Theorem 3: Let $C\binom{N}{j}$ is the number of combinations of N different things taken j at a time and $C\binom{M}{j}$ is the number of combinations of M different things taken j at a time, then the difference $C\binom{N}{j} - C\binom{M}{j}$ can be expressed as

$$C\binom{N}{j} - C\binom{M}{j} = \sum_{i=M+1}^{N} C\binom{i-1}{j-1} \; ; \; j \leq M \leq N \qquad (8)$$

Proof: We get from theorem 2

$$C\binom{N}{j} = \sum_{i=j}^{N} C\binom{i-1}{j-1} = \sum_{i=j}^{M} C\binom{i-1}{j-1} + \sum_{i=M+1}^{N} C\binom{i-1}{j-1}$$

$$= \binom{M}{j} + \sum_{i=M+1}^{N} C\binom{i-1}{j-1}$$

$$\Rightarrow C\binom{N}{j} - C\binom{M}{j} = \sum_{i=M+1}^{N} C\binom{i-1}{j-1}$$

Example 3: Prove the theorem 3 for (i) $C\binom{7}{3}$ and $C\binom{5}{3}$ and (ii) $C\binom{10}{4}$ and $C\binom{6}{4}$.

Solution: We get from (8)

(i) $\binom{7}{3} - C\binom{5}{3} = \sum_{i=6}^{7} C\binom{i-1}{j-1}$

$\Rightarrow 35 - 10 \quad = C\binom{5}{2} + C\binom{6}{2}$

$\Rightarrow 25 \quad\quad\quad = 10 + 15$

$\Rightarrow 25 \quad\quad\quad = 25$

Thus, L. H. S. = R. H. S.

(ii) $\binom{10}{4} - C\binom{6}{4} = \sum_{i=7}^{10} C\binom{i-1}{j-1}$

$\Rightarrow 210 - 15 \quad = C\binom{6}{3} + C\binom{7}{3} + C\binom{8}{3} + C\binom{9}{3}$

$\Rightarrow 195 \quad\quad\quad = 20 + 35 + 56 + 84$

$\Rightarrow 195 \quad\quad\quad = 195$

Thus, L. H. S. = R. H. S.

5. B third combination rule

Theorem 4: The number of combinations of zero different things taken zero at a time would be one i.e.,

$$C\binom{0}{0} = 1 \quad\quad\quad\quad\quad\quad\quad\quad\quad (9)$$

Proof: Consider the Nth term of a Biswas series is

$\sum_{j=1}^{M} V_j C\binom{N-1}{j-1}$

Now putting $N = 1$ and $M = 1$ we get the first term of the Biswas series as

1st term $= \sum_{j=1}^{1} V_1 C\binom{1-1}{1-1}$

$\Rightarrow V_1 = V_1 C\binom{0}{0}$

$\Rightarrow V_1 = V_1 \times 1 = V_1$

i.e., $C\binom{0}{0}$ must be 1.

Another proof can be given by theorem 2 for $j = N$. We have

$$C\binom{N}{j} = \sum_{i=j}^{N} C\binom{i-1}{j-1}$$

Putting $j = N$ we get

$$C\binom{N}{N} = \sum_{i=N}^{N} C\binom{i-1}{N-1}$$

$$\Rightarrow C\binom{N}{N} = C\binom{N-1}{N-1}$$

Now taking $N = 1$ we get

$$C\binom{1}{1} = C\binom{0}{0} = 1$$

i.e., $C\binom{0}{0}$ must be 1.

6. B fourth combination rule

Theorem 5: The number of combinations of N different things taken j at a time where $N < j$ would be zero i.e.,

$$C\binom{N}{j} = 0 \; ; \; N < j \hspace{4cm} (10)$$

Proof: Consider the Biswas series B_M^N. We get first term as

$$1^{st} \text{ term} = \sum_{j=1}^{M} V_j C\binom{0}{j-1}$$

i.e., $V_1 = V_1 C\binom{0}{0} + V_2 C\binom{0}{1} + V_3 C\binom{0}{2} + \ldots + V_M C\binom{0}{M-1}$

$= V_1 \times 1 + V_2 \times 0 + V_3 \times 0 + \ldots + V_M \times 0$

$= V_1$

Thus $C\binom{0}{1}, C\binom{0}{2}, \ldots, C\binom{0}{M-1}$ must be equal to zero. Again for the second term we get

$$2^{nd} \text{ term} = \sum_{j=1}^{M} V_j C\binom{1}{j-1}$$

i.e., $V_1 + V_2 = V_1 C\binom{1}{0} + V_2 C\binom{1}{1} + V_3 C\binom{1}{2} + \ldots + V_M C\binom{1}{M-1}$

$= V_1 \times 1 + V_2 \times 1 + V_3 \times 0 + \ldots + V_M \times 0$

$= V_1 + V_2$

Thus $C\binom{1}{2}$, $C\binom{1}{3}$, ..., $C\binom{1}{M-1}$ must be equal to zero. Now for the third term we get

3rd term = $\sum_{j=1}^{M} V_j C\binom{2}{j-1}$

i.e., $V_1 + 2V_2 + V_3 = V_1 C\binom{2}{0} + V_2 C\binom{2}{1} + V_3 C\binom{2}{2} + V_4 C\binom{2}{3} + ...$
$+ V_M C\binom{2}{M-1}$
$= V_1 \times 1 + V_2 \times 2 + V_3 \times 1 + V_4 \times 0 + ... + V_M \times 0$
$= V_1 + 2V_2 + V_3$

Thus $C\binom{2}{3}$, $C\binom{2}{4}$, ..., $C\binom{2}{M-1}$ must be equal to 3ero. Hence we get
$C\binom{N}{j} = 0$; $N < j$

Example 4: Consider a Biswas series of 3 steps with 5 terms where $V_1 = \frac{1}{2}$, $V_2 = \frac{1}{4}$ and $V_3 = \frac{1}{5}$. Now find the first term applying N^{th} term.

Solution: We have the Nth term as

N^{th} term = $\sum_{j=1}^{M} V_j C\binom{N-1}{j-1}$

Now first term = $\sum_{j=1}^{3} V_j C\binom{1-1}{j-1}$

$= V_1 C\binom{0}{0} + V_2 C\binom{0}{1} + V_3 C\binom{0}{2}$
$= V_1 \times 1 + V_2 \times 0 + V_3 \times 0$
$= V_1$
$= \frac{1}{2}$

Theorem 6: The sum of a Biswas series of M steps with N terms denoted by B_M^N is given by

$B_M^N = \sum_{i=1}^{N} \sum_{j=1}^{M} V_j C\binom{i-1}{j-1}$
$= \sum_{j=1}^{M} V_j C\binom{N}{j}$ ———————————— (11)

Proof: We are now term our attention in series (6)

$$B_M^N = V_1 + \sum_{j=1}^{2} V_j C\binom{1}{j-1} + \sum_{j=1}^{3} V_j C\binom{2}{j-1} + \ldots + \sum_{j=1}^{M} V_j C\binom{M-1}{j-1}$$

$$+ \ldots + \sum_{j=1}^{M} V_j C\binom{N-1}{j-1}$$

$$= V_1 + V_1 C\binom{1}{0} + V_2 C\binom{1}{1} + V_1 C\binom{2}{0} + V_2 C\binom{2}{1} + V_3 C\binom{2}{2} + \ldots$$

$$+ V_1 C\binom{M-1}{0} + V_2 C\binom{M-1}{1} + \ldots + V_M C\binom{M-1}{M-1} + \ldots$$

$$+ V_1 C\binom{N-1}{0} + V_2 C\binom{N-1}{1} + \ldots + V_M C\binom{N-1}{M-1}$$

$$= V_1 C\binom{0}{0} + V_1 C\binom{1}{0} + V_1 C\binom{2}{0} + \ldots + V_1 C\binom{M-1}{0} + \ldots$$

$$+ V_1 C\binom{N-1}{0} + V_2 C\binom{1}{1} + V_2 C\binom{2}{1} + \ldots + V_2 C\binom{M-1}{1} + \ldots$$

$$+ V_2 C\binom{N-1}{1} + \ldots + V_M C\binom{M-1}{M-1} + \ldots + V_M C\binom{N-1}{M-1}$$

$$= V_1 \sum_{i=0}^{N-1} C\binom{i}{0} + V_2 \sum_{i=1}^{N-1} C\binom{i}{1} + \ldots + V_M \sum_{i=M-1}^{N-1} C\binom{i}{M-1}$$

$$= \sum_{j=1}^{M} V_j \sum_{i=j-1}^{N-1} C\binom{i}{j-1}$$

Now taking the B combination series

$$C\binom{N}{j} = \sum_{i=j-1}^{N-1} C\binom{i}{j-1} \text{ we get}$$

$$B_M^N = \sum_{j=1}^{M} V_j C\binom{N}{j}$$

That is

$$B_M^N = \sum_{i=1}^{N} \sum_{j=1}^{M} V_j C\binom{i-1}{j-1}$$

$$= \sum_{j=1}^{M} V_j C\binom{N}{j}$$

Example 5: Let $V_1 = 2$, $V_2 = \frac{2}{3}$, $V_3 = \frac{3}{4}$ and $V_4 = 5$; then find the sums of the Biswas series B_4^{10} and B_4^{12}.

Solution: From eq. (11) we get

$$B_4^{10} = \sum_{j=1}^{4} V_j C\binom{10}{j}$$

$$= V_1 C\binom{10}{1} + V_2 C\binom{10}{2} + V_3 C\binom{10}{3} + V_4 C\binom{10}{4}$$

$$= 2 \times 10 + \frac{2}{3} \times 45 + \frac{3}{4} \times 120 + 5 \times 210$$

$$= 20 + 30 + 90 + 1050$$
$$= 1190.$$

and $B_4^{12} = \sum_{j=1}^{4} V_j C\binom{12}{j}$

$$= V_1 C\binom{12}{1} + V_2 C\binom{12}{2} + V_3 C\binom{12}{3} + V_4 C\binom{12}{4}$$
$$= 2 \times 12 + \frac{2}{3} \times 66 + \frac{3}{4} \times 220 + 5 \times 495$$
$$= 24 + 44 + 165 + 2475$$
$$= 2708.$$

Theorem 7: The Biswas series of M steps with N terms denoted by B_M^N can be expressed as

$$B_M^N = B_M^M + \sum_{i=M+1}^{N} \sum_{j=1}^{M} V_j C\binom{i-1}{j-1} \qquad\qquad (12)$$

Proof: We have from (11)

$$B_M^N = \sum_{j=1}^{M} V_j C\binom{N}{j}$$

Using B combination series we get

$$B_M^N = \sum_{j=1}^{M} V_j \left\{ C\binom{j-1}{j-1} + C\binom{j}{j-1} + C\binom{j+1}{j-1} + \cdots + C\binom{N-1}{j-1} \right\}$$

$$= \sum_{j=1}^{M} V_j C\binom{j-1}{j-1} + \sum_{j=1}^{M} V_j C\binom{j}{j-1} + \sum_{j=1}^{M} V_j C\binom{j+1}{j-1} + \cdots$$
$$+ \sum_{j=1}^{M} V_j C\binom{N-1}{j-1}$$

$$= \sum_{j=1}^{M} V_j C\binom{j-1}{j-1} + \sum_{j=1}^{M} V_j C\binom{j}{j-1} + \sum_{j=1}^{M} V_j C\binom{j+1}{j-1} + \cdots$$
$$+ \sum_{j=1}^{M} V_j C\binom{M-1}{j-1} + \sum_{j=1}^{M} V_j C\binom{M}{j-1} + \cdots + \sum_{j=1}^{M} V_j C\binom{N-1}{j-1}$$

$$= \sum_{j=1}^{M} V_j \left\{ C\binom{j-1}{j-1} + C\binom{j}{j-1} + C\binom{j+1}{j-1} + \cdots + C\binom{M-1}{j-1} \right\}$$
$$+ \sum_{j=1}^{M} V_j C\binom{M}{j-1} + \cdots + \sum_{j=1}^{M} V_j C\binom{N-1}{j-1}$$

$$= \sum_{j=1}^{M} V_j \sum_{i=j-1}^{M-1} C\binom{i}{j-1} + \sum_{i=M+1}^{N} \sum_{j=1}^{M} V_j C\binom{i-1}{j-1}$$

Taking the B combination series we get

$$B_M^N = \sum_{j=1}^{M} V_j C\binom{M}{j} + \sum_{i=M+1}^{N} \sum_{j=1}^{M} V_j C\binom{i-1}{j-1}$$

$$= B_M^M + \sum_{i=M+1}^{N} \sum_{j=1}^{M} V_j C \binom{i-1}{j-1}$$

$= B_M^M +$ the sum of last (N−M) terms of the B series.

Example 6: Consider example 5. Then prove the theorem 7
Solution: We get

$$B_4^{10} = B_4^4 + \sum_{i=5}^{10} \sum_{j=1}^{4} V_j C \binom{i-1}{j-1}$$

Now, $B_4^4 = \sum_{j=1}^{4} V_j C \binom{4}{j}$

$$= V_1 C \binom{4}{1} + V_2 C \binom{4}{2} + V_3 C \binom{4}{3} + V_4 C \binom{4}{4}$$

$$= 2 \times 4 + \frac{2}{3} \times 6 + \frac{3}{4} \times 4 + 5 \times 1$$

$$= 8 + 4 + 3 + 5$$

$$= 20.$$

And, $\sum_{i=5}^{10} \sum_{j=1}^{4} V_j C \binom{i-1}{j-1}$

$$= \sum_{j=1}^{4} V_j C \binom{4}{j-1} + \sum_{j=1}^{4} V_j C \binom{5}{j-1} + \sum_{j=1}^{4} V_j C \binom{6}{j-1}$$

$$+ \sum_{j=1}^{4} V_j C \binom{7}{j-1} + \sum_{j=1}^{4} V_j C \binom{8}{j-1} + \sum_{j=1}^{4} V_j C \binom{9}{j-1}$$

$$= V_1 C \binom{4}{0} + V_2 C \binom{4}{1} + V_3 C \binom{4}{2} + V_4 C \binom{4}{3} + V_1 C \binom{5}{0} + V_2 C \binom{5}{1} + V_3 C \binom{5}{2} +$$

$$V_4 C \binom{5}{3} + V_1 C \binom{6}{0} + V_2 C \binom{6}{1} + V_3 C \binom{6}{2} + + V_4 C \binom{6}{3} + V_1 C \binom{7}{0} + V_2 C \binom{7}{1} +$$

$$V_3 C \binom{7}{2} + + V_4 C \binom{7}{3} + V_1 C \binom{8}{0} + V_2 C \binom{8}{1} + V_3 C \binom{8}{2} + + V_4 C \binom{8}{3}$$

$$+ V_1 C \binom{9}{0} + V_2 C \binom{9}{1} + V_3 C \binom{9}{2} + V_4 C \binom{9}{3}$$

$$= 2 \times 1 + \frac{2}{3} \times 4 + \frac{3}{4} \times 6 + 5 \times 4 + 2 \times 1 + \frac{2}{3} \times 5 + \frac{3}{4} \times 10$$

$$+ 5 \times 10 + 2 \times 1 + \frac{2}{3} \times 6 + \frac{3}{4} \times 15 + 5 \times 20 + 2 \times 1 + \frac{2}{3} \times 7$$

$$+ \frac{3}{4} \times 21 + 5 \times 35 + 2 \times 1 + \frac{2}{3} \times 8 + \frac{3}{4} \times 28 + 5 \times 56$$

$$+ 2 \times 1 + \frac{2}{3} \times 9 + \frac{3}{4} \times 36 + 5 \times 84$$

$$= 2(1+1+1+1+1+1) + \frac{2}{3}(4+5+6+7+8+9) + \frac{3}{4}(6+10+15+21+28+36)$$

$$+ 5(4+10+20+35+56+84)$$

$$= 12 + \frac{2}{3} \times 39 + \frac{3}{4} 4 \times 116 + 5 \times 209$$

$= 12 + 26 + 87 + 1045 = 1170.$

Thus, $B_4^{10} = B_4^4 + \sum_{i=5}^{10} \sum_{j=1}^{4} V_j C \binom{i-1}{j-1} = 20 + 1170 = 1190.$ (Proved)

Again, $B_4^{12} = B_4^4 + \sum_{i=5}^{12} \sum_{j=1}^{4} V_j C \binom{i-1}{j-1}$

Now, $B_4^4 = 20$

And, $\sum_{i=5}^{12} \sum_{j=1}^{4} V_j C \binom{i-1}{j-1}$

$= \sum_{j=1}^{4} V_j C \binom{4}{j-1} + \sum_{j=1}^{4} V_j C \binom{5}{j-1} + \sum_{j=1}^{4} V_j C \binom{6}{j-1}$

$+ \sum_{j=1}^{4} V_j C \binom{7}{j-1} + \sum_{j=1}^{4} V_j C \binom{8}{j-1} + \sum_{j=1}^{4} V_j C \binom{9}{j-1}$

$+ \sum_{j=1}^{4} V_j C \binom{10}{j-1} + \sum_{j=1}^{4} V_j C \binom{11}{j-1}$

$= 1170 + \sum_{j=1}^{4} V_j C \binom{10}{j-1} + \sum_{j=1}^{4} V_j C \binom{11}{j-1}$

$= 1170 + V_1 C \binom{10}{0} + V_2 C \binom{10}{1} + V_3 C \binom{10}{2} + V_4 C \binom{10}{3}$

$\qquad + V_1 C \binom{11}{0} + V_2 C \binom{11}{1} + V_3 C \binom{11}{2} + V_4 C \binom{11}{3}$

$= 1170 + 2 \times 1 + \frac{2}{3} \times 10 + \frac{3}{4} \times 45 + 5 \times 120 + 2 \times 1$

$\qquad + \frac{2}{3} \times 11 + \frac{3}{4} \times 55 + 5 \times 165$

$= 1170 + 2(1+1) + \frac{2}{3}(10 + 11) + \frac{3}{4}(45 + 55) + 5(120 + 165)$

$= 1170 + 4 + 14 + 75 + 1425$

$= 2688.$

Thus, $B_4^{12} = B_4^4 + \sum_{i=5}^{12} \sum_{j=1}^{4} V_j C \binom{i-1}{j-1} = 20 + 2688 = 2708.$ (Proved)

Corollary 1: The Biswas series of M steps with N terms denoted by B_M^N satisfies the following equation

$$B_M^N - B_M^M = \sum_{i=M+1}^{N} \sum_{j=1}^{M} V_j C \binom{i-1}{j-1} \qquad \text{(13)}$$

Theorem 8: Let k is any integer then the Biswas series of M steps with $N \pm k$ terms; $N \pm k \geq M$; denoted by $B_M^{N \pm k}$ can be expressed as

$$B_M^{N \pm k} = B_M^M + \sum_{i=M+1}^{N \pm k} \sum_{j=1}^{M} V_j C \binom{i-1}{j-1} \qquad \text{(14)}$$

Proof: Suppose $N \pm k = N'$ where $N \pm k \geq M$ then from theorem 7 we get

$$B_M^{N'} = B_M^M + \sum_{i=M+1}^{N'} \sum_{j=1}^{M} V_j C\binom{i-1}{j-1} \qquad (15)$$

Putting the value of $N' = N \pm k$ we get

$$B_M^{N \pm k} = B_M^M + \sum_{i=M+1}^{N \pm k} \sum_{j=1}^{M} V_j C\binom{i-1}{j-1}$$

Hence the proof.

Example 4.7: Let $N = 10$, $k = 2$, $V_1 = 2$, $V_2 = \frac{2}{3}$, $V_3 = \frac{3}{4}$ and $V_4 = 5$, then prove the theorem 8.

Solution: We have from example 6

$$B_4^{10} = B_4^4 + \sum_{i=5}^{10} \sum_{j=1}^{4} V_j C\binom{i-1}{j-1}$$

And, $\quad B_4^{12} = B_4^4 + \sum_{i=5}^{12} \sum_{j=1}^{4} V_j C\binom{i-1}{j-1}$

$$\Rightarrow B_4^{10+2} = B_4^4 + \sum_{i=5}^{10+2} \sum_{j=1}^{4} V_j C\binom{i-1}{j-1}$$

Corollary 2: Let k is any integer then the Biswas series of M steps with N $\pm k$; $N \pm k \geq M$; terms denoted by $B_M^{N \pm k}$ satisfies the following equation

$$B_M^{N \pm k} - B_M^M = \sum_{i=M+1}^{N \pm k} \sum_{j=1}^{M} V_j C\binom{i-1}{j-1} \qquad (15)$$

Theorem 9: Let B_M^N is a Biswas series of M steps with N terms and $B_{M_2}^N$ and $B_{M_1}^N$; $M_2 > M_1$ are two Biswas series of first M_2 steps with N terms and second M_1 steps with N terms respectively then $B_{M_2}^N - B_{M_1}^N$ can be expressed as

$$B_{M_2}^N - B_{M_1}^N = \sum_{i=M_1+1}^{N} \sum_{j=M_1+1}^{M_2} V_j C\binom{i-1}{j-1} \qquad (16)$$

Proof: We consider i^{th} terms of $B_{M_2}^N$ and $B_{M_1}^N$; $i < M_1 < M_2$. Now the subtraction of ith terms is

$$\sum_{j=1}^{M_2} V_j C\binom{i-1}{j-1} - \sum_{j=1}^{M_1} V_j C\binom{i-1}{j-1}$$

$$= V_1 C\binom{i-1}{0} + V_2 C\binom{i-1}{1} + \ldots + V_{M_2} C\binom{i-1}{M_2-1} - V_1 C\binom{i-1}{0}$$

$$- V_2 C\binom{i-1}{1} - \ldots - V_{M_1} C\binom{i-1}{M_1-1}$$

We have $M_1 < M_2$ then let $M_2 = M_1 + k$. So we get the ith term of $B_{M_2}^N - B_{M_1}^N$ as

$$= V_1 C\binom{i-1}{0} + V_2 C\binom{i-1}{1} + \ldots + V_{M_1} C\binom{i-1}{M_1-1} + V_{M_1+1} C\binom{i-1}{M_1}$$
$$+ \ldots + V_{M_1+k} C\binom{i-1}{M_1+k-1} - V_1 C\binom{i-1}{0} - V_2 C\binom{i-1}{1} - \ldots$$
$$- V_{M_1} C\binom{i-1}{M_1-1} \hspace{4cm} (17)$$
$$= V_{M_1+1} C\binom{i-1}{M_1} + V_{M_1+2} C\binom{i-1}{M_1+1} + \cdots + V_{M_1+k} C\binom{i-1}{M_1+k-1}$$
$$= \sum_{j=M_1+1}^{M_1+k} V_j C\binom{i-1}{j-1}$$
$$= \sum_{j=M_1+1}^{M_2} V_j C\binom{i-1}{j-1} \hspace{4cm} (4.18)$$

Now from (17) the first M_1 terms of the subtraction result zero, so i goes from M_1+1 to N i.e., the sum over i of (18) is as

$$\sum_{i=M_1+1}^{N} \sum_{j=M_1+1}^{M_2} V_j C\binom{i-1}{j-1}$$

Hence we get

$$B_{M_2}^N - B_{M_1}^N = \sum_{i=M_1+1}^{N} \sum_{j=M_1+1}^{M_2} V_j C\binom{i-1}{j-1}$$

Example 8: Let $V_1 = 2$, $V_2 = 3$, $V_3 = 5$ and $V_4 = 6$ and $N = 6$. Then find (i) $B_4^6 - B_1^6$,
(ii) $B_3^6 - B_2^6$ and (iii) $B_4^6 - B_2^6$.

Solution: We have

(i) $B_4^6 - B_1^6 = \sum_{i=2}^{6} \sum_{j=2}^{4} V_j C\binom{i-1}{j-1}$

$$= \sum_{i=2}^{6} \left\{ V_2 C\binom{i-1}{1} + V_3 C\binom{i-1}{2} + V_4 C\binom{i-1}{3} \right\}$$
$$= \left\{ V_2 C\binom{1}{1} + V_3 C\binom{1}{2} + V_4 C\binom{1}{3} \right\} + \left\{ V_2 C\binom{2}{1} + V_3 C\binom{2}{2} + \right.$$

V4C23 + V2C31+V3C32+ V4C33 + V2C41+V3C42+ V4C43 +

$$\left\{ V_2 C\binom{5}{1} + V_3 C\binom{5}{2} + V_4 C\binom{5}{3} \right\}$$

Using $C\binom{N}{M} = 0$; $N < M$, we get

$B_4^6 - B_1^6 = \{3 \times 1 + 5 \times 0 + 6 \times 0\} + \{3 \times 2 + 5 \times 1 + 6 \times 0\} + \{3 \times 3 + 5 \times 3 + 6 \times 1\}$

$$+ \{3\times4 + 5\times6 + 6\times4\} + \{3\times5 + 5\times10 + 6\times10\}$$
$$= 3 + 11 + 30 + 66 + 125$$

Now,

$$B_4^6 = 2 + 5 + 13 + 32 + 68 + 127$$
$$B_1^6 = 2 + 2 + 2 + 2 + 2 + 2$$

So, $B_4^6 - B_1^6 = 3 + 11 + 30 + 66 + 125$

(ii) $B_3^6 - B_2^6 = \sum_{i=3}^{6} \sum_{j=3}^{3} V_j C\binom{i-1}{j-1} = \sum_{i=3}^{6} \{V_3 C\binom{i-1}{2}\}$

$$= V_3 C\binom{2}{2} + V_3 C\binom{3}{2} + V_3 C\binom{4}{2} + V_3 C\binom{5}{2}$$
$$= 5\times1 + 5\times3 + 5\times6 + 5\times10$$
$$= 5 + 15 + 30 + 50$$

Now,

$$B_3^6 = 2 + 5 + 13 + 26 + 44 + 67$$
$$B_2^6 = 2 + 5 + 8 + 11 + 14 + 17$$

So, $B_3^6 - B_2^6 = 5 + 15 + 30 + 50$

(iii) $B_4^6 - B_2^6 = \sum_{i=3}^{6} \sum_{j=3}^{4} V_j C\binom{i-1}{j-1}$

$$= \sum_{i=3}^{6} \{V_3 C\binom{i-1}{2} + V_4 C\binom{i-1}{3}\}$$
$$= \{V_3 C\binom{2}{2} + V_4 C\binom{2}{3}\} + \{V_3 C\binom{3}{2} + V_4 C\binom{3}{3}\}$$
$$+ \{V_3 C\binom{4}{2} + V_4 C\binom{4}{3}\} + \{V_3 C\binom{5}{2} + V_4 C\binom{5}{3}\}$$

Using $C\binom{N}{M} = 0$; $N < M$, we get

$$B_4^6 - B_2^6 = \{5\times1 + 6\times0\} + \{5\times3 + 6\times1\} + \{5\times6 + 6\times4\} + \{5\times10 + 6\times10\}$$
$$= 5 + 21 + 54 + 110$$

Now,

$$B_4^6 = 2 + 5 + 13 + 32 + 68 + 127$$
$$B_2^6 = 2 + 5 + 8 + 11 + 14 + 17$$

So, $B_4^6 - B_2^6 = 5 + 21 + 54 + 110$

7. Conclusions

The series used in discrete computations in arithmatics. Combination theorems widely used in this paper.

References

1. Deapon Biswas, Paper 6, Summation methods, Bystematics My Classic, 2010 Self published, Chittagong, 2016 Monon Prokashon, Chittagong, Bystematics Vol. I, My Classic, 2018 Scholar's Press EU, ISBN: 987- 620-2-30664-5.

2. Deapon Biswas, Paper 13, On the combinations, Bystematics My Classic, 2010 Self published, Chittagong, 2016 Monon Prokashon, Chittagong, Bystematics Vol. II, My Classic, 2018 Scholar's Press EU, ISBN: 987- 620-2-30960-8.

3. Deapon Biswas, Paper 15, B series, Bystematics My Classic, 2010 Self published, Chittagong, 2016 Monon Prokashon, Chittagong, Bystematics Vol. II, My Classic, 2018 Scholar's Press EU, ISBN: 987- 620-2-30960-8.

4. F. Mosteller, R. E. K Rourke & G. B. Thomas Jr. Probability with Statistical Applications.

5. Carl B. Allendoerfer & Cletus 0. Oakley. Principles of Mathematics.

YOUR KNOWLEDGE HAS VALUE

- We will publish your bachelor's and master's thesis, essays and papers

- Your own eBook and book - sold worldwide in all relevant shops

- Earn money with each sale

Upload your text at www.GRIN.com and publish for free